INTERMEDIATE TECHNOLOGY
Myson House, Railway Terrace
Rugby CV21 3HT

USING TECHNICAL SKILLS IN COMMUNITY DEVELOPMENT

USING TECHNICAL SKILLS IN COMMUNITY DEVELOPMENT

An analysis of VSO's experience

by Jonathan Dawson

Edited by Mog Ball

Practical
ACTION
PUBLISHING

VSO/IT Publications 1990

Practical Action Publishing Ltd
25 Albert Street, Rugby,
Warwickshire, CV21 2SD, UK
www.practicalactionpublishing.com

A catalogue record for this book is available from the British Library & Library of Congress

ISBN 978-1-85339-078-4 Paperback
ISBN 978-1-78044-374-4 Digital book

Citation: Dawson, J. (1990) *Using Technical Skills in Community Development: An analysis of VSOs experience*, Rugby, UK: Practical Action Publishing https://doi.org/10.3362/9781780443744

Since 1974, Practical Action Publishing has published and disseminated books and information in support of international development work throughout the world. All print editions are produced and distributed via ethical and sustainable print on demand global facilities.

Practical Action Publishing is a trading name of Practical Action Publishing Ltd (Company Reg. No. 01159018 | VAT 880 9924 76). All profits are covenanted back to its parent group, Practical Action (Charity Reg. No. 247257).

The manufacturer's authorised representative in the EU for product safety is Lightning Source France, 1 Av. Johannes Gutenberg, 78310 Maurepas, France. compliance@lightningsource.fr

CONTENTS

FOREWORD

Anyone concerned with rural development, and especially with technologies appropriate to the needs and resources of the rural poor, will find these seven case studies of absorbing and topical interest. Between them, they cover a wide spectrum of the problems encountered by practical fieldworkers committed to the task of helping people to help themselves, by bringing within their reach, and ownership, the tools and equipment with which they can work themselves out of their poverty.

Twenty-five years ago, when we started the Intermediate Technology Group, we believed that the central objective was to discover and make known the right kind of hardware – tools and equipment that were small, simple and inexpensive enough for the rural poor to own and operate. Experience has shown that the hardware only works if it is part of a package which empowers local people to choose what suits them best; gives them access to a low-cost, good quality product over which they have control; and which enables them, by using it, to raise their standard of living on a sustainable basis. Essentially this is a process of investing in people by making them more productive.

This does not mean that the technologist cannot do a good job unless he or she is also an expert anthropologist, extension and community development specialist. But it does mean that the technologist must be sensitive to the social, cultural and economic environment into which the technology must fit if it is to serve its purpose. Take time to learn and don't try to make big leaps are two of the lessons drawn from these case studies that appeal to me; but there are others, relating to training, project preparation and teamwork that are relevant not only to VSO but to all agencies working at grassroots level – and increasingly such organizations are being formed within the developing countries themselves.

There is now a growing recognition that the large-scale technologies of conventional aid and development programmes have little or nothing to offer the majority of people in developing countries. I am certainly not alone in believing that the future lies with technologies appropriate to the needs and skills of men and women in the rural areas and small communities of poor countries. I commend these case studies as a contribution to the challenging task of discovering what the rural poor are trying to do, and helping them to do it better.

George McRobie

A founder member,
and now Vice-President of the
Intermediate Technology Development Group

1. INTRODUCTION

'I feel envious when I look at volunteers working in formal institutions. The structures in which they work, the curricula they teach and the objectives they are seeking to achieve are fixed. By comparison, to those of us working in less structured posts, it is often unclear what is the most effective way of working and, ultimately, we frequently have difficulty in identifying just what it is precisely we are trying to achieve.'

These are the words of a water engineer who worked as a VSO volunteer on a community-based programme. They will be familiar to many volunteers whose task it is to share their technical skills in non-institutional settings in developing countries. Such volunteers often have to struggle, both to define their role, and to identify the most appropriate and efficient way of using and disseminating their skills. This study seeks to analyse the experience of a number of projects involving technically skilled workers working in non-institutional settings (i.e. with informal groups or communities). It looks at the work the volunteers undertook, the problems they faced, their responses to those problems and the wider lessons to be drawn from their experience.

The experience can perhaps best be introduced and described in general terms as follows. A volunteer, who might be a water engineer, a forester, an agriculturalist, a craft worker, or a civil or mechanical engineer, arrives in post. The task involves working directly with people (individuals, groups, or communities) to enable them, in the long-term, better to help themselves. The volunteer is usually a well-qualified, proficient technician, with previous work experience. Yet shortly after arriving it is realised that technical skills alone are not going to be sufficient for the job in hand. The volunteer has to embark on a process of cultural adaptation and self-education in a range of non-technical skills, and in the most appropriate way to apply those technical skills that they already possess.

Central to this learning and adaptation process is the need to

Language training for newly arrived volunteers. Up to six months spent on cultural familiarization pays off for technical volunteers who are involved in the early stages of community or group-based projects.
Photo: Kate Seed

understand the importance of familiarization with the cultural environment within which skills transfer takes place. In addition, in the design of projects, the importance of community participation at all levels and stages also needs to be understood, as does the need for the volunteer to possess a wide range of skills other than specifically technical ones.

There are several groups of such skills which expatriate technical volunteers need to have. The first can be broadly defined as 'community work' skills. If it is assumed that volunteers are the 'experts', who have exclusive access to the 'solution' of the problem, they will have no more to do than to deliver that solution and leave. If, on the other hand, the long-term solution is seen to be in the hands of the community itself, in whose service the technician works, then the community must develop the skills, and the volunteer's task is to build this capacity.

To do this, the volunteer, the volunteer agency, and the local requesting agency must all ensure that the work arises out of the needs and aspirations of the community. Volunteers then also need to be prepared and versatile enough to address meetings, give formal or informal training, do extension work, and identify social as well as technical hurdles to skills dissemination. This means that volunteers have to possess or learn skills which will often run contrary to their professional instincts. In particular they need to have the patience to help *others* to define problems and overcome them, rather than do such things *themselves*.

A second group of skills is closely related. Sometimes, technical volunteers are working not with communities, but with more discrete formal or informal groups – perhaps a co-operative, or a women's group. In these circumstances, volunteers need a good understanding of how such groups function, and of how to deal with such group-dynamic issues and processes as conflict resolution, decision-making, problem-solving, and the like. When the volunteers are seen as, or cast themselves into the role of, the group leader (and therefore themselves act as conflict resolvers, decision-makers and problem-solvers), short-term gains are made at the expense of long-term development.

Thirdly, it is in the nature of the tasks undertaken by technical volunteers that they find, in the course of their work, that they need some quite specific skills and capabilities which they do not possess. The most obvious example is that in which a volunteer helping a group or community to make, say, bricks from local materials realizes that marketing and business planning and management skills are needed if the bricks are to be sold.

Fourthly, technical volunteers often find themselves needing to train or teach others their skills, as part of the process of transfer to local ownership and control. But not every technician is a natural or skilled teacher or trainer.

Finally, and running across all the previous four, technical volunteers need the ability to strike a balance between *doing* and *facilitating*, in every aspect of their work. Volunteers, motivated strongly by a desire to *do*, often find it hard achieving this balance, for it means doing *less* and facilitating others to do *more*, against the objective of building the capacity of others to the point where the volunteers are no longer needed.

All of this has many implications: for volunteer training, for the process by which projects are identified, and for the ways in which volunteers work and are supported.

This study examines VSO's experience in the field of the deployment of technically skilled volunteers, by examining the experience of volunteers in seven diverse projects.

However, because of the wide variety of projects in which VSO participates and the range of countries in which it does so, the case studies each offer experience and lessons, both positive and negative, which appear to be representative of many other projects in which volunteers and agencies are working with communities or groups. The problems faced and the potential solutions are of general interest and concern to all those involved in the effective use of technical expertise in community-based projects.

Liz Osborn confers with Ihssen Abdul Raheem on the pricing of her products, in Sudan. Marketing skills are among the range of additional skills which technical volunteers often find that they need but do not have.
Photo: Jeremy Hartley

2. SUMMARY CONCLUSIONS

The experience outlined in the seven case studies which comprise the following chapter describes the impact of technically skilled volunteers on the needs of groups and communities in the Third World. Part 4 then presents an analysis of the experience, which indicates that the impact of the volunteers is likely to be maximized when the efforts of the volunteers themselves, of those agencies requesting them, and of those agencies recruiting, preparing, training and posting them, produce the following conditions.

PREPARATION

Requesting agencies need to undertake thorough initial *preparatory and development work* in order to maximize the volunteer impact. In particular, where local communities or other target groups are to be both beneficiaries of, and play their part in the establishment and longer-term ownership of, the project, their awareness must be raised and efforts mobilized well before the Volunteer arrives in post. The closer requesting agencies are to the beneficiary groups and communities, and the better known to and trusted by them they are, the more likely it is that the optimum conditions will be created. Whilst volunteers, when in post, can and should play their part in group and community awareness-raising, (as they will be able to when they are well prepared through prior training and cultural familiarization), they should complement and continue previous efforts and not be seen as a substitute for them.

Requesting and volunteer-sending agencies should make great efforts to *analyse the skill and experience requirements* of the posts to be filled by the volunteers. Such analysis should enable the agencies to identify in detail the skills and experience volunteers will need to have beyond their specific technical capabilities and qualifications.

This analysis should then inform volunteer *selection* and both pre-departure and in-country volunteer cultural/technical/non-technical *training and familiarization* activities. Many projects will demand, ideally, either a multi-skilled volunteer and/or a team approach. But the more that the individual volunteers have flexible, versatile dispositions, based upon a prior realisation of the range of activities to be undertaken and the range of skills required, the better. Technical volunteer training should in part at least be orientated to developing such flexible and versatile attitudes.

TIMING

Requesting and volunteer-sending agencies also need jointly to make astute judgements as to the *timing* of volunteer postings. It is important that the local conditions described above have been created before a volunteer is posted.

VOLUNTEER WORK AND SKILLS

Technical volunteers involved in the early stages of community or group-based projects need to spend longer periods on *cultural familiarization* at the beginning of their postings than those working in institutionalized settings. A period of up to six months spent on this would appear to be appropriate. Volunteers should not be placed under, nor create for themselves, pressure to achieve practical results and outcomes during this period.

Volunteers should in general regard the *transfer* of their skills as just as important – if not more so – than the *application* of them. To achieve this, prior training in, and/or on-the-job learning of, the skills of group work, facilitation, and teaching/training will be highly important. Skills transfer is also more likely to take place in circumstances where volunteers work as members of teams comprised of locally recruited volunteers or paid workers.

As long-term projects proceed, and as volunteers are replaced by others, requesting and volunteer-sending agencies need constantly to reappraise the skill and experience requirements of the projects, and to select volunteers accordingly. It should be recognized that as projects develop, the skill and experiences that need to be brought to them by volunteers will change.

VOLUNTEER SUPPORT

The varied and demanding nature of the work, the range of skills and experience required by it, and the changes that both the job and the skills/experience go through over time, mean that even the most versatile volunteers will be likely to require higher levels of both local and external support and guidance than those working in institutionalized settings.

3. SEVEN CASE STUDIES

Joe Gomme, water engineer, Nepal recognized that the success of his work could not be measured in terms of the number of pumps installed but by the extent to which local communities were able to maintain them, and the degree to which the health benefits of using clean water were conveyed.
Photo: Sue Eckstein

NEPAL: THE COMMUNITY WATER SUPPLY PROGRAMME

THE PROJECT

Since 1979, Nepal's Community Water Supply (CWS) programme (run by the then Ministry of Panchayat and Local Development) has had the aim of supplying water of reasonable
quality to every village in Nepal by the year 2000, through dig-
ging wells and constructing tap-stands. The building of latrines
has also been added to the programme's activities. The programme is centrally planned, with the Ministry in Kathmandu
deciding on the timing of work in each district and on the number
of facilities to be installed.

VOLUNTEER INPUT

VSO has been involved since the inception of CWS, with more than
20 volunteers having worked for the programme as water engineers. One of the more recent volunteers is Joe Gomme, who has
an MSc. in Hydrogeology and had worked for an engineering consultancy firm in Britain before leaving for Nepal.

Joe worked for CWS as an engineer/overseer in charge of a
team of technicians. These technicians were responsible for super-
vizing the installation work carried out by private contractors. Joe's
task was to provide the technicians with training in both the technical aspects of their work and in promoting community participation.
He was also responsible for overseeing and helping to train a team
of women assigned to do primary health care education work for
the programme in his district.

CWS programme-implementation in each area in Joe's district
would usually take the following form. Two months would elapse
between the arrival of the CWS team in the area and the beginning
of construction work. During this period, the team performed
two principal functions. The first (which proved generally

unproblematical) was concerned with technical matters – carrying out a water resource and population survey and, in co-operation with the villagers, selecting sites for water sources. Only a modest level of technical skill was required to carry out these tasks. A previous volunteer who had worked with the programme described this aspect of the work as 'technically unsatisfying, as only a few weeks are required to comprehend what is involved'.

The second function of the team was that of communicating to the communities the importance of developing local maintenance facilities, and awareness of health issues.

During the two years he was in post 400 hand-pumps were installed (providing water sources for 6,000 households), 400 latrines were produced and sold, and 100 technicians and female education workers received training.

LESSONS

Joe feels, however, that the success of the work cannot be measured by such statistics. For him, the two key measures of success were: the degree to which a maintenance ability for the installations was developed in the community; and the successful communication of the health benefits to be derived from clean water. Both of these elements of success were dependent on the extent to which the voluntary mobilisation and participation of the targeted communities could be achieved.

It was not easy to achieve success on these fronts. The CWS programme as a whole had, after all, come not from the communities themselves but from the Ministry in Kathmandu, and Joe found that in most cases, while some leading members of the community knew of the initiative, most of the villagers did not.

Joe had to an extent been sensitized to the cultural environment in which he was to work by his first six months in Nepal. The first two months were taken up with language training in Kathmandu. He then lived and worked in a village for four months before taking on his responsibility for his CWS programme district. He felt that the village experience was of great importance in giving him an insight into 'community dynamics' and in preparing him, in some measure, for the community mobilization and participation in which he would be involved. Nevertheless, he felt poorly prepared for some of the training aspects of his work. The communication of technical information to the technicians created few difficulties. It was much harder to present ideas and information to the villagers in a form which was relevant and accessible, whether this was done directly by Joe in the community meetings which launched the programme in new areas or, indirectly, through training the technicians and education workers to do the same.

Joe had no experience whatsoever of playing a training or teaching role. He feels that, as a result, he was too impatient, tending to do things himself rather than facilitate others to define problems and find answers. Despite the fact that he was charged with the training of the team of health education workers, he had no knowledge of primary health care issues. In a number of important respects, then, Joe felt he lacked certain skills which were critical to the success of the programme and was ill-prepared to communicate what knowledge he did possess.

One blatant example cited by Joe of his failure to appreciate the importance of local, cultural factors was what he described as 'the inability of most illiterate Nepali villagers to understand ideas presented in pictures – simply because they are completely unused to the concept that real things can be represented on paper.'

While it is tempting to measure the success of projects such as the CWS programme in terms of the number of handpumps and latrines installed, using the wider measure of the long-term sustainability of the installations and the health spin-offs arising from clean water is considerably more difficult but, ultimately, more meaningful. If the second measure is to show success, too, it is clear that the volunteers involved need a wider range of skills than those encompassed by their engineering and technical capabilities and qualifications.

Tenant farmers in the Philippines had little incentive to transfer the innovations developed with agriculturalist Liz Kiff on the communal farm to their own plots because the benefits would accrue to their landlords – a cultural fact that had not been sufficiently appreciated.
Photo: Jenny Matthews

THE PHILIPPINES: THE PAGLAUM
COMMUNITY-BASED DEVELOPMENT AGENCY
(PACOBDA)

THE PROJECT

PACOBDA started in 1976 as an integrated agricultural develop-ment programme involving health, cooperatives and farming in the Damulog district of the Bukidon province of Mindanao. By 1983, when VSO volunteer agriculturalist Liz Kiff arrived to work with PACOBDA, it consisted of a team of six community workers acting principally as facilitators in helping farmers set up co-operative stores and maize-marketing organizations. Under the guidance of the volunteer the objective was to rekindle the programme's objec-tives of improving agricultural practices and extending the range of productive activities.

VOLUNTEER INPUT

Liz Kiff had a BSc. in Agricultural Botany, while studying for which she spent a year working for the Ministry of Agriculture. Her original VSO posting had been to an agricultural college as a lecturer in agronomy, but, for a variety of reasons, this proved unsuitable. In seeking to find employment elsewhere, she got in touch with PACOBDA (which already had links with VSO, having employed two volunteers some years before). It was agreed that Liz would work for PACOBDA to relaunch its agricultural programme.

Liz arrived to find that few people other than the team of com-munity workers knew exactly why she had come. She had to start from scratch, deciding where she was going to be based, with which communities she was going to work, and how she was going to promote the idea of increased and diversified agricultural pro-duction. A mammoth task for any worker, it was all the more so for a young, inexperienced person recently arrived in what was for her an alien culture.

Liz travelled around the area covered by PACOBDA talking to people about agricultural problems. Eventually she chose, on a rather arbitrary basis, one of the communities to be her operational base. Together with a group of interested farmers she organized a visit to a nearby agricultural training centre, which had the effect of generating considerable interest.

Then, two co-operative groups with which she spent much of her time secured a loan for a three-hectare farm. While the farm had to be profit-making in order to repay the loan, its primary function was as an area on which to experiment and demonstrate new ideas for improved techniques and crops. Liz hoped that successful innovations would then be transferred to the farmers' private farms.

The experiments and innovations were various. A vegetable plot was started which introduced many new crop varieties. Terracing and tree planting were introduced to demonstrate their usefulness in combatting soil erosion. The benefits of intercropping were demonstrated. A fishpond was dug and stocked. Grain storage techniques were developed. As all this went on there were encouraging signs of community participation: members of the two co-operative groups provided all the labour for the farm and some modest success was achieved in transferring some of the innovations onto the farmers' private plots.

Within a year of Liz's departure, however, the farm had failed and little remained thereafter of the idea and innovations generated during her stay. Why was this the case?

LESSONS

The initiative clearly arose not out of the perceived needs of the community for self-improvement, but of the volunteer for employment. While the participation of the co-operative groups in the farm suggests the project was at least to an extent meeting a local need, it was probably not given a very high priority by the groups themselves. The dominant role was played by the volunteer; the very survival of the farm was largely dependent on Liz playing such a role. Yet while PACOBDA had a great deal of community animation experience which could have allowed Liz to play both a more supportive and technical assistance role, she was, for the most part, left to manage all aspects of the work herself.

In short, Liz felt unskilled in, and unprepared for, *either* the leadership role she played *or* for the more supportive role which a better relationship with PACOBDA would have produced.

She also felt untrained in keeping the farm's records and accounts, a task for which she took sole responsibility. Nor did she know how to prepare for her departure by transferring this function to a community member.

Local land-ownership patterns also undermined the success of the agricultural programme. Most of the farmers with whom Liz was working were tenants on the land of estate-owners. As a result, there was little incentive for them to transfer the innovations developed on the communal farm onto their own plots, for the benefits would simply have accrued to their landlords. This cultural fact was insufficiently appreciated by the volunteer, yet it was clearly a highly influential one.

Perhaps at the heart of all these lessons was the decision to launch the project without the need for it having been clearly identified by the targeted beneficiaries themselves.

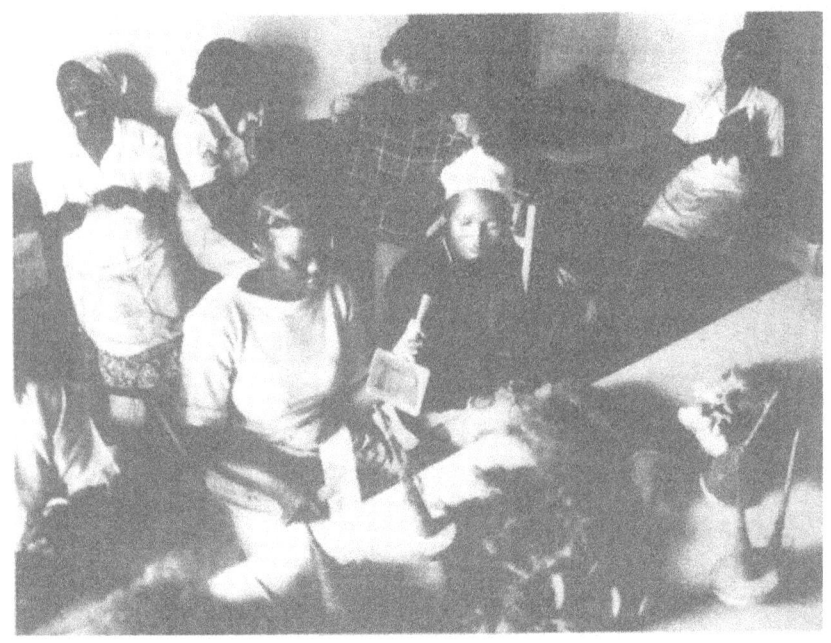

Members of the Manyoya women's knitting group in Kenya with business adviser Patricia Sellers. Patricia succeeded Helen Derbyshire who had provided technical training to the group. Both needed skills for working within a group and appreciated the importance of taking a back-seat advisory role rather than being directive.
Photo: Stephen Pern

KENYA: THE MANYOYA WOMEN'S KNITTING GROUP

THE PROJECT

The Manyoya women's knitting group was started in 1984. It was one of a number of income-generating initiatives begun in the early 1980s under the aegis of the Shamberere Rural Education Programme, the object of which was to address the problem of high levels of unemployment among school-leavers in the South Kabras district of Kenya's Western Province through the provision of technical and business training.

The Manyoya group consisted of seven women. Their initial task was to raise their technical ability to the point where they could make their products and then find a market for them.

VOLUNTEER INPUT

VSO volunteer Helen Derbyshire arrived in Shamberere before the launching of Manyoya. A social worker with several years' experience, she had originally been sent to Kenya as a food nutritionist working with another group operating under the Rural Education Programme. But she had already done some part-time work for the Manyoya group and now transferred to it on a full-time basis in late 1984.

Initially, Helen's input was exclusively technical, teaching the women to spin, to hand-knit, and to improve their machine-knitting skills. The success of this training was evidenced by contracts being won by the group from two prestigious Nairobi shops, following an exhibition of their products at national craft fairs. Indeed by late 1985 the group was in the fortunate position of being unable to meet the ever-growing demand for its products.

In September 1985, one of the women opened her own workshop in the local market, employing and training her own workforce. During the following year, the five other surviving members of the original group followed suit, and they steadily expanded to the point where they employed a total of approximately 70 women.

By this time, with little further need for technical training, Helen found herself having to play an ever-wider role, one for which she had received no formal training, but on whose success the continued development of the Manyoya group was in large measure dependent. The problem, as she saw it, was that the Manyoya women considered Helen to be 'the boss'. They lacked the confidence, despite Helen's prompting, to play a larger role in the group's management. Helen continued to make most of the decisions – about designs, marketing, pricing, etc. Such decisions, Helen felt, should be made by the women themselves.

Armed with this understanding of the problem and with a high degree of sensitivity to the needs of the women, she set out to do as well as she could. Aware of the danger of creating dependency on herself and of further undermining the women's self-confidence, she refused the offer of personal motorized transport. She always travelled with the women, on foot or by public transport. She was careful to give credibility and support to the original group of women, now employers in their own right. She also followed a deliberate policy of belittling her own abilities (in which, she claims, little self-modesty was required!) and of always underlining the women's own strengths.

The process by which the women began to exercise a greater degree of control over their own group happened very slowly. On a marketing trip to Nairobi, on which, as far as was possible, Helen played a 'back-seat' role, the signs of her own weaknesses relative to the Manyoya women, in hawking and bargaining, played an important part in encouraging them to reappraise their own worth and ability.

As a result, ideas and initiatives slowly began to arise from within the group. They established themselves as a trade association, opened and administered their own bank account and began to make marketing trips without Helen. While all were agreed that Helen should be replaced on her departure by another VSO volunteer, it is a measure of the enhanced self-confidence of the women that they saw that the need was for a volunteer with different skills, in identifying new markets and in developing the group's ability to manage their business.

So, in early 1987, VSO volunteer Patricia Sellers, a business manager with six years' experience working as a store manager and as branch manager of a recruitment consultancy firm, arrived in Shamberere. After one year in post, she was able to report that each of the women had complete control over their work groups, doing all of the accounts, paying wages, controlling stock, doing quality control and organising marketing and sales. By the end of her two-year contract, new markets had also been opened up.

LESSONS

Despite all these signs of evident success, in which the positive lessons are clear, Patricia feels in retrospect that in certain respects she was poorly equipped to tackle the job in hand. Like Helen, she found that proficiency in her professional specialization was not, in itself, sufficient. Firstly, for example, the problems presented by occasional unharmonious group dynamics and by personal attitudes among the women towards business practices taxed her. She found their attitudes quite different from anything she had previously experienced, largely because they were of a cultural rather than a technical nature.

Both volunteers felt frustrated in slowly guiding others towards a goal when their professional instincts were rather to do things quickly themselves. But the fact that the Manyoya group now operates without any external assistance is, in large measure, due to the conscious efforts of both volunteers to play the role of facilitator rather than achiever: this is the major lesson.

Andy Bennett, community builder with the Muka Mukuu Society in eastern Kenya, 'relearned his entire approach to building in the Third World' by visiting building projects, talking to Kenyan builders and researching materials for six months before beginning any construction work.
Photo: Stephen Pern

KENYA: THE MUKA MUKUU LOW-COST BUILDING PROGRAMME

THE PROJECT

Muka Mukuu covers 11,560 hectares in Kenya's Eastern Province. Its principal products are coffee and sisal; the sales of both are organised through a co-operative association, the Muka Mukuu Society. The population has grown steadily in recent years to a current estimated 17,000. As the population has increased, so there has been a growing strain on housing stocks.

In direct response to the United Nations International Year of Shelter for the Homeless, the Muka Mukuu Society requested assistance from VSO to help it establish a low-cost housing project in the area.

The primary aim of the project was to improve the standard of basic housing by introducing new construction methods, developing existing building techniques and maximizing the use of locally available materials. By thus lowering the cost of such buildings, the hope was that they would become more affordable for individual members of the community, and thus achieve the second aim, of increasing the housing stock. The process of development and transfer was to take the form, firstly, of providing on-site training to teams of local volunteers building 12 'project houses', and then secondly, to promote the establishment of small, construction-related enterprises to build others.

These aims were to be achieved through the construction of one 'project house' as a residential unit at each of the ten primary schools in the area, and two more at the area's polytechnic. The participation of the local communities in all aspects of their local project house – siting, design, provision of labour, etc. – was a central aspect of the project design.

VOLUNTEER INPUT

In early 1987, a request was made for three volunteers – a project co-ordinator/building manager, a community worker/social surveyor and a general builder. In the autumn of that year, the first two of these jobs were filled by two volunteers. Ann Martin was a qualified community education worker and had worked in community education on the Isle of Skye. Andy Bennett had an Higher National Diploma in engineering and an MSc. in Business Studies. He began his working life in 1966 as a civil engineer and had managed his own building company.

VSO was, however, initially unable to find a volunteer to undertake the third job, of general builder. Partly as a result of this, and partly because of their own ideas on the best way to approach the work, Andy and Ann decided to adopt a 'slowly, slowly' approach and to engage in a substantial exercise of educating themselves, and of beginning to mobilize community interest and participation, before embarking on the physical construction work.

They did a considerable amount of travelling during this initial period, which lasted six months. They visited other building projects, talked to Kenyan builders and studied their methods. Andy researched locally available materials, markets and prices. Ann, meanwhile, worked with the communities resident in the localities of each of the ten schools, helping them to establish building committees and to define their needs and objectives. She also channelled their ideas to Andy.

Thus it was not until six months after their arrival, having been joined by VSO volunteer Chris Granger (who had worked for 25 years as a self-employed builder) that construction work began on one of the sites. This initial period was felt to be far from wasted. As Andy put it: 'For the first six months at least, I was shelving many of my preconceptions and relearning not just new methods, but my entire approach to building in the Third World.'

It is still too early to pass judgement on the success or otherwise which has been achieved. There are, however, encouraging signs that this lengthy process of education and development may be beginning to bear fruit. The level of community participation in the building work has been very encouraging, with local volunteers providing all the labour. This is all the more impressive in view of the hardship caused locally by two successive years of below average rainfall.

Another positive indication lies in the fact that a number of the local volunteers have already established independent building work using materials and techniques introduced by the VSO volunteers. A tile production unit established by Andy has now been transferred to local management and the project's blockmaking is now contracted to local builders. The eagerness with which the new ideas and techniques have been adopted provides further evidence that Andy's research into the technical aspects of the work has been of considerable value.

LESSONS

Thus, positive lessons about the approach were already emerging in 1989. Others can be drawn from the fact that three different tasks/skills were identified as being needed and were (albeit by accident) supplied by VSO in two stages. Ann, the community worker, arrived a full six months before the first brick was laid. Even so, she feels that she should have been introduced even earlier. In particular, she feels the decision, taken before her arrival by the leaders of the society, to allocate the 'project houses' to the schools as residential units, is one which she found to be against the wishes of the communities. This made her work more difficult. (In fact, this decision was largely dictated by Ministry of Education insistence on the availability of accommodation in advance of the allocation of teachers.) Ann found that the communities themselves would have preferred to see the units housing community services, such as clinics, and acting as community group meeting places, etc. (It was, in fact, subsequently agreed that a number of the houses would serve both residential and community functions.)

The timing of the arrival of Andy, the co-ordinator, appears to have been near perfect, even if his initial research period was, to some extent, enforced by the unforeseen event of the absence initially of the third volunteer. As it turned out this was fortuitous. The builder's skills would not have been relevant to the needs of the situation in the autumn of 1987: had they been available then, the project might not have turned out so well.

Different problems – and therefore lessons – lie ahead. The biggest concerns the mechanisms by which control of the project will be localised. In addition, there may be a need for a further change in the composition of the skills provided by VSO, should small, construction-related enterprises develop, as is hoped.

It's easier as a woman to gain the trust and respect of women and involve them in community activities. Jacinta Barrins was conscious of this when working with women on the construction of wells and water tanks in Kenya. Nearly two-thirds of her work was, broadly, 'community work'. The picture shows women's group co-ordinator Nicky May with two members of the Chekalini Women's Group.
Photo: Jeremy Hartley

KENYA: THE KITUI COMMUNITY-BASED WATER DEVELOPMENT PROGRAMME

THE PROJECT

In late 1986, VSO's Kenya field office was approached by the Development Committee of the Diocese of Kitui, in the east of the country. Following several years of below average rainfall and a rapidly growing local population, the diocesan committee had established a programme to improve the access of rural communities to water sources. The situation was already severe, with many of the seasonal rivers having dried up completely and women in some areas having to walk up to 40 kilometres to fetch and carry home water.

VOLUNTEER INPUT

In February 1987, VSO volunteer Jacinta Barrins arrived in Kitui to work as a water technician. A civil engineer with a diploma in water engineering, she had worked on commercial urban drainage projects and in the Department of Engineering Hydrology at University College, Galway. Her job in Kitui was to provide technical advice to participating communities on such matters as the most appropriate form of water catchment and the siting of the wells, dams and rock catchments.

Much work had, however, been done before her arrival. Parish priests, in liaison with their parishioners, had identified a considerable number of local groups prepared to contribute their labour and funds. A large number of possible sites had also been identified by these groups. The diocesan committee had secured financial backing from five donor agencies, appointed two area co-ordinators (to be responsible for liaising between the water technician and participating groups) and had purchased a vehicle for use by the volunteer. It had also set out forward plans which contained budgetary provisions for foreseeable future needs such as the employment of artisans to lead the construction teams on each of the sites.

The importance of the preparatory work done before Jacinta's arrival cannot be overstated. It was done through the long-established, well-respected diocesan structure, which was in continual contact with its communities, thus representing for them a secure and well-known partner whose commitment and competence could be counted on, and to which communities could feel confident about giving much in return, including labour, food for workers, accommodation (for artisans brought in from outside) and a proportion of the costs of the cement.

There were thus mass community participation, well-organised groups with elected committees and the two area co-ordinators trained in community mobilisation, all in place before the volunteer arrived.

Jacinta, then, represented the final piece of a carefully planned and constructed jigsaw. On arrival, she was given a list of 135 sites proposed by community groups. On her departure, two years later, the groups with which she had been working had completed the construction of 21 dams, 22 shallow wells, five rock catchments and 267 water tanks. A large number of other sites were in various stages of completion.

LESSONS

For Jacinta, all this preparatory work did not mean that she just did a technical job. Her first six months represented a re-education period during which she made a number of mistakes by, in her own words, 'doing too much talking and not enough listening', and 'trying to push things too fast'. Had she continued in the same vein, trying to dictate the pace of the work to suit her own needs, she feels that there would have been a real risk of alienating many of the participating groups from their own projects.

Sensitive to this danger, Jacinta found she had to learn to play a more withdrawn role, responding to the groups' efforts rather than trying to initiate them. This involved gaining the trust and respect of those she was attempting to serve, becoming involved in community activities, and listening and observing. She feels it was much easer to achieve this as a woman. Local women (who comprised the bulk of the labour force for the construction work) would never have been able to communicate so freely with a man, she says.

In addition, she was sensitive to the need to delegate authority progressively, to discourage dependence on her. This involved some informal training of the two area co-ordinators to make them more able to perform the technical aspects of the work. It also involved a conscious effort to make the groups themselves, and particularly their committees, more assertive and self-confident.

Jacinta estimates that only about 35 per cent of her work could strictly be described as being purely *technical*. The other 65 per cent was broadly 'community work' in nature. The technical work presented no problems. The community work, on the other hand, was one for which she had little personal or professional training. That she was able to learn it on-the-job is the result of the thorough preparations made before her arrival and of her own sensitivity and adaptability.

How to convince farmers that they would have to shell their maize immediately on harvesting was one of the problems faced by agriculturalist Caroline Hanks in a project to control the larger grain borer in Tanzania. Caroline found that communication skills were as vital to her work as her technical knowledge.
Photo: Jane Mason

TANZANIA: LARGER GRAIN BORER CONTROL PROGRAMME

THE PROJECT

In the late 1970s, the larger grain borer was accidentally introduced into the Tabora region in central Tanzania. It later spread to all but three of the nation's regions, and has had a devastating effect on stored maize and cassava, two of the country's main food crops, damaging 70–80 per cent of maize cobs for example. In 1981, the British Overseas Development Administration provided funding and some personnel to the Ministry of Agriculture in Tanzania in an attempt to find an antidote to the borer.

By 1983 research indicated that small-scale use of a chemical product which had previously been used on a large scale on South American plantations might have some beneficial effect. The project's aim was to encourage its use by local farmers and to make such refinements to the chemical mix as were appropriate in the light of emerging experience.

VOLUNTEER INPUT

VSO volunteer Caroline Hanks was posted to the research station in Tanzania which was undertaking the tests. She had recently graduated with an MSc. in the Technology of Crop Protection.

Caroline had been in post for about a year when the station declared itself satisfied with the suitability of the chemical for conditions in Tabora and asked her to take responsibility for getting farmers to use it. It now remained only for local farmers to accept the chemical and the techniques for using it. At first sight it might appear that most farmers would need little incentive to change their practices as they had already suffered much from the borer. However, its acceptance proved much more difficult than anticipated, for a number of reasons.

First, farmers were discouraged by their experience three years earlier with another chemical issued to combat the borer with the approval of the Ministry of Agriculture. This had been introduced without sufficient research, and had proved both costly and ineffective.

Secondly, the new techniques involved a major change in post-harvest work practices which initially proved unacceptable to many. The borer only attacked maize on the cob, and the new techniques therefore involved shelling all the maize immediately after the harvest, rather than in small quantities for two or three days' use. Farmers were resistant to making the change.

Thirdly, the agency which provided sacks in which the shelled maize was to be stored grossly underestimated the quantities that would be needed.

Finally, Caroline had neither the experience nor the training to enable her speedily to identify and tackle all these emerging problems.

Essentially, she felt that she lacked the necessary communication skills. She felt she had little idea how to go about communicating the message with which she had been charged or understanding why it was meeting with such resistance. This was a cultural problem.

Nevertheless her major contribution was in encouraging the manufacture and use of large baskets from locally available materials for shelled maize storage. This was never an easy task and required her to exert continual pressure on the basket makers.

VSO continued its participation in the Grain Borer Control Programme and in 1990 had eight volunteers working cn it in various parts of Tanzania. The focus of the volunteers' wo k is very firmly on extension activities aimed at promoting the adoption of the new storage techniques. The objective of placing volunteers, as spelled out in their job description, is, 'to motivate, support and train the extension officers, and ultimately the farmers'. Despite the fact that the need for specialist agricultural expertise is much reduced from that required during Caroline's first year, all of the volunteers in post in 1990 have agricultural/scientific backgrounds.

LESSONS

A recently returned volunteer feels that rather more weight ought now to be laid on the specific skills required to do *extension* work. 'To be really effective,' she said, 'an extension worker must under-

stand what is happening in the community, what to attach importance to or to ignore.' The greatest skills need she felt she had was
for 'communication and cultural understanding'. It was only by the
end of her two-year contract that, 'the pieces were coming together', and she began to feel confident in this vital area. In short,
without communication skills and cultural knowledge, agricultural
skills and knowledge, in this project, were wasted.

What future for this seed dresser at Tikonko Agricultural Extension Centre in Sierra Leone? Design engineer Paul Watkins discovered a 'museum' of technologies that had been developed by previous volunteers, but that for the most part had not been adopted because of a lack of research, business management and extension work.

SIERRA LEONE: THE APPROPRIATE TECHNOLOGY WORKSHOP, BO TIKONKO

THE PROJECT

The Tikonko Agricultural Extension Centre began operating in 1969 under the direction of a committee made up of representatives of the Methodist Church, the local community and the Ministry of Agriculture. In 1979, following a policy review, the decision was taken to broaden the scope of the Centre's activities. Rather than acting purely as a service unit for the Centre's own vehicles and machinery, it was to play a more active role in developing and disseminating technologies appropriate to improving local agricultural practices and output.

VOLUNTEER INPUT

VSO has had a long history of involvement with the Centre, providing volunteers to work as agriculturalists for the farm and engineers for the appropriate technology workshop. The most recent volunteer was Paul Watkins. Paul had qualified with a Higher National Diploma in mechanical engineering from Bristol Polytechnic and had worked for five years as a design engineer.

On his arrival Paul was surprised that even 'with such an impressive record of volunteer co-operation' he could find 'so little for me to study or follow-up from previous volunteers'. The legacy of the previous volunteers was what he described as a 'museum' of technologies which they had developed but which for the most part had been adopted neither by farmers nor artisans.

No structured research had been undertaken into local technologies or agricultural practices. The project management was not over-directive and it was expected that Paul would take the initiative and play a leading role in determining which projects were to be developed.

'It took me up to six months to discover where best to start,' he said. While he had all the technical skills necessary to develop technologies relevant to the needs of local farmers and artisans, he felt unable fully to utilize those skills, partly because of the lack of research done by the Centre before his arrival and partly because of his own limitations in non-technical fields.

He felt ill-equipped, for example, to conduct research to gauge local technical and technological capacity, even though such research was clearly essential for him to gear his innovations to local needs.

He also lacked even the most basic business skills, skills which would have enabled him to cost his new technologies to ensure that they would be affordable. While the Centre could use its overseas contacts to import components and materials not otherwise available, or at much reduced prices, the danger was that the product could not then be imitated by local artisans at similarly low cost or, worse still, these artisans might be undercut by the Centre competing against them, while having the advantage of artificially subsidized materials.

Paul, by his own admission, was both professionally unprepared and personally disinclined to do as much research into local society and technologies as he saw, in retrospect, to be necessary. His project manager noted that 'he didn't see much relationship between technology, culture and society', and that, 'he wanted to prescribe the technology to answer local problems and hoped people would accept it, instead of the other way around'.

The Centre itself also failed to provide this knowledge and Paul increasingly retreated into the technical and technological world with which he was familiar – a world which, in most cases, bore little relation to that inhabited by local artisans and farmers.

LESSONS

The lessons are clear enough. At the end of his contract in 1986, Paul listed the various projects on which he had worked, describing the success which each had enjoyed. Through the list run two threads. The first tells a story of continual technical enhancement. The second describes, with few exceptions, the failure of local producers to adopt the innovations. 'A small wood-turning lathe was constructed and is now at the workshop in Tikonko. It performs well but has not really caught the imagination of the carpenters in the rural areas,' he reported. Inappropriate design, unavailability of materials and excessive expense were repeatedly cited as reasons for the failure to transfer technology out of the workshop, as they were for other innovations such as cassava graters, cassava presses, culvert moulds, ox-carts and vices for shaping timber.

Yet the real reasons lay elsewhere, in failings of cultural understanding and communication. Paul himself seemed to recognize this. He said: 'It is becoming more and more difficult to design anything radically new that the farmer can afford to take a risk with.'

4. THE EXPERIENCE ANALYSED

The chart on the following page attempts to summarize each of the seven case studies under a number of key descriptive features. The closer projects are to achieving // or / ratings under each feature, the closer they are to what might be termed the 'optimum' state, where volunteer impact is maximized. This chapter develops the analysis shown in the table.

THE CONTEXT IN WHICH VOLUNTEERS WORK

If projects are to maximize the impact and performance of technical volunteers, it is clear they must see those volunteers as one part of a much larger picture, a part to be put in place not just the right way, but at the right time, too. The place and time will differ from project to project. But what about the other parts? The first and most important of these is the need for a high level of participation by the beneficiaries at as many stages of projects as possible, from conception to completion. Yet, only in some of the cases examined is such continuing involvement apparent. As regards project conception and design, for example, a number of the projects developed from actions taken by external agencies. Commonly, as with the water supply programme in Nepal, it is a government ministry which plans and implements projects. They are then, nevertheless, still dependent on community participation. In the case of the appropriate technology workshop in Tikonko, the idea also came, it would appear, from outside the community, from a group of well-intentioned people. But here, there appears to have been little attempt to mobilise community participation subsequently by the volunteer. Similarly, in the case of the PACOBDA agricultural programme, the initiative had little to do with the perceived needs of the community, and much more to do with the designs of the volunteer. While attempts to mobilise the subsequent participation of the beneficiary groups were made, they had little effect.

In contrast, in the cases of Kitui, Manyoya and Muka Mukuu, the

Overall analysis of the seven case studies

Project	Source of the project idea : (Local = //; External = /; from vol = x)	Extent of preparatory/development work before vol. arrival	Extent of preparatory/development work by the volunteer	Extent of volunteer cultural familiarisation	Volunteer alone (x) or part of local staff team (/)	Work needing tech/prof. skills beyond main qualific. W	P	Social and community work W	P	Group work W	P	Facilitation W	P	Training/teaching W	P
CWS Nepal	/	nk	//	//	/	na	na	/	x	//	x	/	x	//	x
PACOBDA Philippines	x	x	x	nk	x	/	x	/	x	na	x	x	x	x	x
Manyoya Kenya	//	/	nk	/	x	//	x	na	na	//	x	//	/	//	x
Muka Mukuu Kenya	/	//	//	//	x	/	x	//	/	//	/	na	na	na	na
Kitui Water Kenya	//	//	//	nk	/	/	x	//	x	nk	nk	//	x	//	x
Grain Borer Tanzania	/	x	x	/	x	/	x	x	x	/	x	x	x	x	x
Tikonko Sierra Leone	/	x	/	x	x	x	x	x	x	x	x	x	x	x	x

Volunteer willingness (W) and preparedness (P) to undertake

Notes: nk = not known; na = not, or not yet, applicable.

role of community-based or -linked organizations was of central importance in facilitating the mobilisation of the beneficiary groups and communities and in helping them define their objectives. The groundwork done in these cases, in mobilising the beneficiaries, was a major factor in creating the conditions for the technical progress achieved later with the help of the volunteers.

It would appear that this process of 'beneficiary involvement' occurs most naturally, and at an unforced pace, when projects are small-scale and localized. But some of the examples described show that it can also be achieved where the projects are part of national programmes, albeit with more difficulty. It has to be recognized that this is an inherent weakness in large-scale programmes, which take key decisions externally, to 'target' communities or groups for a defined form of assistance. Small-scale, autonomous, community-based projects appear much more likely to be able to allow the beneficiaries to choose both *what* is the most appropriate form of assistance and *how* that assistance should be delivered.

The problems faced by larger-scale, externally motivated programmes, are well illustrated by the case of a fisheries project in West Africa, not described here, in which VSO was participating in 1990. It is a multi-pound project covering six villages, each of which is to be provided with a fishing service centre to promote the volume of fish caught, processed and sold. The administrative requirements of the funders have forced on to the local project management a degree of inflexibility which, in many respects, appears to be incompatible with genuine community-based decision-making. The objectives of the project as well as its timing were established in advance, and specified to those involved.

Volunteers working with such projects are aware of potential conflicts between the communities and project managers. One of them said, 'The real test is yet to come: when I implement a decision made by the fishing community within its own democratic system only to be told by the project manager that the decision is not acceptable because . . .' But even in projects where groups or communities have been 'targetted' for assistance by centralized bureaucracies and subjected to their decisions, much can be done by volunteers and community-based or -linked organizations, to explain the objectives and benefits of the innovations to be introduced by the project. The experience in Muka Mukuu suggests that this work should be taken very seriously, by making a significant amount of time available for community consultation. But in the Community Water Supply Programme in Nepal, the two-month-period in which the team can get their (quite complex) messages across appears to be inadequate.

The most important parts of the picture which should mostly be in place before the arrival of the volunteers but which can be developed by them thereafter relate, then, to the mobilization and education of the beneficiary communities or groups. But in certain types of project, there may be a need for an additional input in the form of research into the social, cultural, economic or technological characteristics of the project communities or groups. At least two of the case studies show that if a volunteer technician is to direct a technical skill towards meeting the perceived needs and problems of artisans and farmers, s/he must have a thorough understanding of their current situation. The consequences of omitting this piece of the picture are most clearly illustrated by the experience of the volunteers at the Tikonko appropriate technology workshop and in the Grain Borer Programme in Tanzania.

What lessons can therefore be drawn from the examples examined about the management of projects? These can be broken down into three parts: involvement of the beneficiaries in project management; the management of the volunteer; and the management of the process of project development.

THE BENEFICIARIES

Here, a number of lessons stand out. The first, which follows from the analysis presented in the previous paragraphs, is that the participation of the beneficiaries in managing projects must be given the highest priority. The pace at which the project proceeds and the manner in which it does so should, wherever possible, be dictated by the beneficiaries. Even in the Kitui water project, where the process had been very carefully planned, the volunteer could have threatened the project's progress, if she had not been sensitive to this need.

The importance of the beneficiaries contributing their labour resources, spirit and enthusiasm recurs repeatedly in the examples selected. Uppermost at all times in the minds of project managers and volunteers should be the question of how and when control can be transferred to them. The case of PACOBDA illustrates the danger of giving insufficient thought to this. This process, however, has to be conducted with great sensitivity and patience: the danger is of removing assistance from communities or groups before they are able to stand on their own feet. Most commonly, as is evident in the case of the Manyoya group, the problem is one of lack of self-confidence among those with whom the volunteer is working.

It is worth noting that while in only two of the examples described (CWS Nepal and Kitui, Kenya) were the volunteers working as members of locally recruited teams of workers, there appears to be value in such a situation, in two respects. Firstly, local workers are, in a sense, an intermediary group, working between the volunteer and the main beneficiary groups, and can play an important role in securing the participation of the beneficiaries. Secondly, the problems associated with skills transfer, dealt with later in this chapter, also become more soluble when the transfer is initially to the members of the working team, rather than to the beneficiaries themselves.

It should be borne in mind that the beneficiaries of volunteer work will often tend to be the poorest and least-educated groups in society. Thus, almost by definition, it will take a great deal of time before they are confident about performing tasks such as keeping accounts, organizing meetings, dealing with ministries and other outside agencies. The development of management capacities among benificiary groups and communities will often, therefore, be a long-term process.

VOLUNTEER MANAGEMENT

As regards the implication for the ways in which volunteers should be managed, two lessons are apparent. In the first place, the temptation to give the volunteers too much scope for their own initiative should be firmly resisted. It should at all times be remembered that volunteers come from quite alien social and material cultures. If they are to work in the service of a community or group, they will need strong direction and support to help them understand just what the needs are and how to go about meeting them.

This point is most clearly illustrated by the Tikonko example. A well-qualified and experienced design engineer, the volunteer was ill-equipped to handle his apparently open-ended brief and would very likely have responded to more directive, or at least more supportive, management.

The second lesson about the management of volunteers concerns the importance of reducing the pressure on them, especially to achieve tangible results during their early months in post. Many volunteers felt, in retrospect, that their first few months would have been best spent listening to and watching local communities, farmers and artisans, not 'achieving' in any sense of the word they would previously have understood, but, nonetheless, laying the necessary social and technical foundations for their future work.

The value of such an introductory period of 'non-achievement' is confirmed by those who did just that, as evidenced by the examples described. The 're-education process' described by Andy Bennett in Muka Mukuu is most striking in this respect, echoed by Joe Gomme in Nepal. All felt that their productivity over the years of their service was higher for having spent time at the beginning redefining personal objectives while understanding local needs and cultures.

PROJECT MANAGEMENT

What do the examples described say about project management? Large-scale, externally defined and motivated programmes tend to have to adhere to prescribed objectives. These will involve the achievement of certain quantifiable aims such as the number of wells or tap-stands installed. Yet, in much community-based work, success can often only be measured in *qualitative* terms, by for example measuring the use which is made of an innovation or the local capability to maintain and nurture whatever equipment or development is involved. The point is made by a number of volunteers, especially by Joe Gomme in Nepal.

Another volunteer had this to say on the subject: 'In community work the volunteers' achievements may be difficult to quantify or recognise, as they need to keep a low profile and encourage and enable others in the community. Often their "success" may not be known for some years.'

VOLUNTEER SKILLS

Volunteers are generally most comfortable and proficient when using the technical skills in which they are professionally qualified.

All of the volunteers in the examples described were highly skilled and many had a substantial amount of pre-volunteer work experience. But a number found the work to be technically unfulfilling and unchallenging, feeling that it could have been done by less highly trained technicians.

While there can be little doubt about the volunteers' technical abilities, it is legitimate to question the direct relevance of these abilities to the conditions in which they work.

In Tikonko, for example, while being highly proficient in his own right, the volunteer was simply not able to turn his skills and ideas into innovations relevant to, and within the grasp of, local farmers and artisans. His problems were exacerbated by a failure to give sufficient weight to socio-cultural factors.

By contrast Andy Bennett, the building engineer/co-ordinator with Muka Mukuu, described the benefits of his six months' foundation building' period in socio-cultural and technical terms. 'The dawn of potential technical possibilities is just illuminating my thoughts,' he said.

There is thus often the need for volunteers to be able to adapt their technical skills to different conditions, particularly among agriculturalists. Liz Kiff, working with PACOBDA in the Philippines, was one of a number of agricultural and forestry volunteers who said she had to learn about local conditions before her British-learned knowledge was of real value to the people with whom she was working. Similarly, another VSO agriculturalist observed that 'as far as agriculture is concerned, I am sure I have as much to learn from the farmers as they have from me'.

Even if there is often a need to refine and adapt volunteers' technical skills to local conditions, it is nonetheless in their use and application that they feel most proficient and relaxed. There are, however, other skill areas, which can loosely be defined as 'community work', 'group work', 'extension', 'facilitation and training', and 'other technical and professional', in which volunteers are often not proficient.

It is not easy to define which of these various groups of skills volunteers tend to need most. The examples studied show that the answer varies according to circumstances, and may also vary as time passes, and as projects develop and mature. What can be said with more certainty is that it is possible, through good preparatory work by requesting and volunteer-sending agencies, to determine what skills, beyond the technical ones, volunteers will be likely to need in their posts. It can also be said with some certainty that what *many* technical volunteers will need is the ability to recognize that there is a social dimension to the way in which skills transfer takes place; that technology and skills are not culturally neutral; and that their dissemination is dependent on their being socially as well as technically appropriate. All of this has been stated so often in recent years as to make it a central orthodoxy in the field of development work. Yet the examples studied here reveal that the message is not always getting across.

The example of Tikonko is the clearest one of where the work has been technology-led rather than needs-led. The volunteer appeared to be only vaguely aware of the central importance of the social context within which he was working, and there was similar low awareness among most of the volunteers interviewed in connection with this study. Even in the case of the Kitui water scheme, despite the extensive animation work which had been done prior to her arrival, and the emphasis given to the social environment in her job description, the volunteer's initial failure to work within parameters established by the communities themselves could have threatened the project's success.

'EXTENSION' SKILLS

Understanding the social dimension is a necessary prerequisite to understanding the need for and ways to apply what have been termed community work, group work, and facilitation and training skills. These skills, which are all concerned with aspects of what is commonly called 'extension' are quite different from technical ones. While they are possessed by very few technical volunteers, a number show themselves, in the examples described, to be creative, versatile, willing and sensitive enough to try to learn and practise these skills while in their posts. One returned volunteer said, 'it was only through an understanding of the way local society worked, both socially and politically, and the way people regarded the issues in the context of their own lives, that I gained insight into matters such as whom to approach, how to approach them, what issues they considered priorities . . .' Another volunteer concluded, on the basis of her experience, that 'the volunteer will need to learn about: power structures; communication routes and mores; how to get necessary resources; identifying key people in the community and who gets on with whom; gender roles; and training for adulthood . . .'

The volunteers who had most evident success in understanding and applying the various 'extension' skills had a number of characteristics in common. Firstly, they tended to have put a lot of time and effort into gaining proficiency in the local language. Secondly, they spent a lot of time, socially as well as professionally, moving with the people among whom they worked. One volunteer in the examples described, it will be noted, considered this to be of such importance to the success of her work that she turned down the offer of her own motorised transport, preferring instead to walk and travel on public transport with the members of the group for which she worked. A final characteristic – though one which by definition can hardly be developed! – is gender. Female volunteers often appear to be better able to communicate at all levels of society than males, particularly in situations – which are common – where a heavy economic and social burden is carried by women, as is often the case in rural areas.

There are two particular 'extension' skills in which volunteers tend to be weak, and on which they could therefore benefit from pre-departure training: these are skills in training and teaching, and the related skills of facilitation (i.e. in broad terms being able to help and support others to develop their skills and capacities in ways other than direct teaching and training, in the context of everyday work).

Learning to play the facilitator role, as opposed to the doer/achiever role, is a challenge for many volunteers. This is hardly surprising, since central to the technician's training and experience, and to the motivation of many volunteers, is the drive for personal achievement. Playing the 'back-seat' role is also difficult for volunteers because of the fact that the *pace* of the work, when dictated by the beneficiaries, will generally be considerably slower than that to which they have become culturally accustomed. In addition, where volunteers are working on large-scale, national programmes, developed by government or other external agencies, there may be pressure to achieve results quickly. Thus in these circumstances, volunteers may well find themselves subject to conflicting demands and pressures. One volunteer refers to 'the temptation of being out in front and being seen to be doing everything', while another points out that 'working closely with the community means that your work, meetings, and eventually, decisions, are dictated by factors like the speed, efficiency, enthusiasm and will of the group, which can cause frustrations'.

Despite such frustrations, the successful volunteer appears to be the one who manages to redefine the role that he or she plays as that of a servant rather than leader, facilitator rather than achiever, delegator rather than executor. 'Too many times in my opinion,' said a technical desk officer working in VSO's London office, 'were stories told of volunteers who had 'got things done' – set up nurseries themselves, wrote forest management plans themselves, chaired local meetings, etc. etc. I got the impression that the (community forestry) volunteers seemed to fall into two categories: those who, despite being personally and professionally frustrated by the "system", still sought to achieve whatever results they could by working through it; and those who chose to opt out of the system and start down the slippery slope of doing things themselves.' There can be no doubt that the achievements of the first category of volunteer, while less spectacular than those of the second, will tend to be more sustainable.

OTHER TECHNICAL AND PROFESSIONAL SKILLS

The final group of skills in which many of the volunteers in the examples described tend to be deficient is comprised of a wide range of specific skills in which the volunteers will generally not have had any training or experience. One such skill – evident in the case of Tikonko – is the ability to do social and economic surveys, which require what might be termed 'research' skills. Many volunteers say that they would have been more effective if they had some elementary training in or knowledge of what can be termed 'business' skills, such as record- and account-keeping, costing, marketing, and budgeting.

There are, then, a wide range of skills in which technical volunteers often find themselves to be poorly equipped. It might be argued that many of these skills, perhaps particularly in the community and group work fields, cannot be taught to volunteers, but only learned through experience. This may be true to a degree, but there does appear to be scope for considerable improvement. One of the volunteers had this to say: 'In general, such skills are not so much overlooked, as considered something that individuals will pick up as a matter of course during their work. To a large extent this is true, but an awful lot of valuable time is wasted whilst acquiring the necessary insight which, to a certain extent, could be passed on to new volunteers.'

What can be argued more certainly is that the ideal, most effective technical volunteer is one who has the benefit of thorough training and preparation (which concentrates on helping them develop the complete range of skills needed on the job), a creative, versatile and flexible disposition, and a willingness to learn. To an extent, creativity and a willingness to learn can substitute for prior training, as some of the examples show, but a good volunteer-sending programme will not rely on this.

SOME IMPLICATIONS

How can VSO, and the other organizations involved in volunteer work, improve the impact and effectiveness of technical volunteers working in community settings in the Third World?

Firstly, turning to project design, volunteers must be seen as part of a much larger picture, and the *timing* of their input should be dependent on the other parts of the picture being in place. Secondly, there must be a recognition that volunteers need to possess a variety of skills. Thirdly, great care needs to be taken to identify them and to identify when, in the development process, they will be needed.

Assuming the skill needs are, through good relationships and preparatory work by requesting agencies and volunteer-sending organizations, so identified, there are a number of ways (which are not mutually exclusive) of dealing with the question of how to ensure, or at least to try to ensure, that volunteers possess them. Firstly, volunteer agencies should attempt to provide training in the required range of skills for each volunteer. Secondly, there can and should be a *sequence* in the skills supplied by successive volunteers (as happened with the Manyoya group for example). Thirdly, as in Muka Mukuu, it may be possible to send a *team* of volunteers.

The last approach appears to be almost the ideal, although it is of course the most expensive and complex. Adding to it a degree of flexibility in the timing of different inputs, together with a recognition that different skills inputs tend to be overlapping rather than strictly sequential, one can perhaps see something resembling the perfect approach. As regards timing and overlapping, the case of Manyoya is instructive. The demand for improved business and marketing skills predated by some months the arrival of the business adviser. Similarly, in all the case studies the need for technical and other skills inputs always ran in parallel, rather than in sequence, to the need for 'extension' skills.

The value of the multi-skilled team is most clearly illustrated by the only example where VSO has thus far experimented with it, in the case of Muka Mukuu. Much of the project's success can be put down to the flexibility in the timing of the inputs (albeit partly achieved by accident) and the complementarity of the different skills provided by the three volunteers.

But, as noted earlier, the multi-skilled team should ideally not just be made up of volunteers: part of it should be (and is in the cases of CWS Nepal, and Kitui, Kenya), made up of nationals of the countries in question, even members of the communities and groups benefitting from the projects.

The development of the multi-skilled team approach is hindered, even prohibited in some cases, by a number of factors. In the first place there is reason to believe that the idea is not favoured by agencies requesting volunteers. Larger agencies, and government ministries in particular, tend to be so obsessed with the quantitative, technical aspects of projects that they only see the need for skilled technicians. On a number of occasions, the presence of Peace Corps volunteers, who tend to be generalists rather than specialists, was cited as a factor which discouraged employers of VSO volunteers from applying for non-technicians.

There are also practical and financial problems. A community may only be able to provide the accommodation and salary for one volunteer, and there is indeed a dilemma here when agencies such as VSO are concerned to target their efforts to serving the poorest of the poor.

VSO's programme in The Gambia was in 1990 engaged in an interesting experiment aimed at overcoming some of these problems. A team of eight volunteers, all recruited by the VSO Business and Social Development recruitment desk, but each having a variety of supplementary skills – such as in horticulture, fisheries, management, etc. – were each responsible for community animation work in one village. In addition, however, they all met from time to time to share their skills with each other, and, by extension, with each of the eight villages. At the time of writing it is too early to see what success has been achieved by this interesting innovation, but it appears to hold some promise.

While a combination of teamwork (in various forms), good timing, and an accurate assessment of skills needs (as a continuing process) represent the ideal, much needs to be done, and can be done, to avoid some of the worst outcomes evidenced by some of the examples described. This book provides conclusions and a range of recommendations on the kinds of actions that should be taken by VSO and others involved in community development work.

THE ECOE PROGRAMME

(Evaluating and Communicating our Overseas Experience)

THE NEED

Over the past thirty years, more than 25,000 volunteers have worked abroad with VSO. Currently, there are over 1,200 volunteers working in over 40 developing countries in Africa, Asia, the Pacific and the Caribbean for periods of two years or more. However, we have become increasingly aware that much of this valuable experience has been lost through not being recorded in ways which make it accessible and communicable. The ECOE Programme addresses this problem.

THE AIM

The aim is to record volunteers' experience in reports, videos, seminars, conferences, books, etc. This body of knowledge supplements and supports the work of individual volunteers. It also provides information which is accessible not only to volunteers but also to their employers overseas and to other agencies for whom the information is relevant. Care is taken to present each area of volunteer experience in the context of current thinking about development so that VSO both contributes to development discussions and learns lessons from them for the continuance of its work.

ADVISORY PANEL

A panel of opinion leaders in relevant professions and in development thinking advises on the selection and commissioning of ECOE publications.

For further information write to:

The Programme Evaluation Manager
VSO
317 Putney Bridge Road
London SW15 2PN, UK

Tel: 081-780 2266   Fax: 081-780 1326